The Reason

By

Ian Beardsley

ISBN: 978-1-312-23800-8

The reason the Gypsy Shaman, or Chovihano, Manuel showed me a collection of 15 rubber hoses; that is that at least I believed there were 15 hoses, I understand now, is that in my experience in the Sacromonte (Gypsy Caves of Granada, Spain) I only heard people play flamenco or deep song in the meters of 3 (fandango), 4 (tangos), 6 (bulerias), 8 (tangos), 9 (18 or six three times being bulerias), 12 (Soleares), and Rumba (16). Other than 2, 7, 11, and 15 all is accounted for as 14 is seven twice. 15 is both an expression of 5 and 10. The two is Manuel's taoistic vision that "Dios es una idea, Son dos" (God is an idea, there are two elements). So, 15 stands outside of guitar as I knew it, and Manuel as shaman, stands outside of flamenco and deep song looking in.

I realize now, too, that since flamenco is marked 12, 3, 6, 10, 12 as accents in twelve beats, and you start counting on twelve in the standard view of the most vibrant genre, called bulerias, that my sequel to my work "Discover And Contact" called "Discover And Contact Two", was suggested because you can start counting on one in threes and get 1, 4, 7, ... for accents.

Thus in my work Discovery And Contact we worked the with the set (9/5, 5/3, 2) which we can call a functional group and we worked with 15 outside the set, which we can call an operator, as I conveniently like to look at it now. I called this the basis of a Cosmology, but perhaps I should be clear that it is not the meaning of Cosmology in Astronomy, which is the scientific approach to understanding the origin and fate of the Universe as well as its structure, but is the term as used in Archaeology, which can refer to any prehistoric, or culture's creation myth that they use to account for the Universe. We worked with the new set in my work Discover And Contact Two that is (1, 4, 7).

We are now left in this work to consider 7 and 11. We only count as far as 16 because that is considered the ultimate form of expression in North Indian Classical Music.

We begin by looking at 7 divided by 11 or the fraction seven elevenths (7/11) which divided out is 0.636363636...

We at once notice it has the repetition of 36 out to infinity, and make the connection that 3.6 is 1.8 multiplied by two and 1.8 is nine-fifths (9/5). Remember that 9/5 was one of three key numbers in the Cosmology we laid out in Discover And Contact. Nine to five is the molar mass of gold to that of silver. Nine to five is the ratio of the solar radius to the lunar orbital radius; The sun is gold in color, the moon silver.

3.6 is the second term in the sequence where each successive term is a whole number multiple of 1.8. Venus is the second planet out from the sun. It is 0.72 astronomical units from the sun (average earth-sun separations). It is considered a failed earth and 7.2 is the fourth term in the sequence where each successive term is a whole number multiple of 1.8. Mars is the fourth planet from the Sun. It is the most promising planet to colonize.

Seven is the planet Uranus, there is no eleventh planet (at least that we know of). The planets are named after Ancient Greek Gods or their Roman Equivalents. Uranus is the Greek God of the sky and husband and son to Gaia, who was the personification of the Earth and was the mother of all. Venus is the Goddess of love, one could say she is the Greek version of the East Semetic Akkadian, Assyrian, or Babylonian Ishtar. It would seem all can be traced back to the Ancient Sumerians, who were the predecessors of these civilizations in Ancient Mesopotamia.

11 diviided by 7 is the comparison of 11 to 7 but, it makes just as much sense to add 7 to 11 because it is the separation between bodies when they are on opposite sides of the sun, one at 11 the other at 7. 7+11 = 18 and 7+11 is like 6 + 12 which is 18. This is the essence of the flamenco genre bulerias, it is 6 three times.

We make the two sequences:

$a\_n = (11/7)1, (11/7)2, (11/7)3, (11/7)4,...$

And,

$a\_n = 7, 7+11=18, 18+11=29, 29+11=40,...$

Which gives,

$a\_n = 1.57, 3.142857, 4.714, 6.2857,...$

And,

$a\_n = 7, 18, 29, 40,...$

We take their difference and get:

$a\_n=5.43, 14.857, 24.286, 33.7143,...$

$14.857-5.43 = 9.427$ $24.286-14.857 = 9.429$ $33.7143-24.286 = 9.4283$

And we see the common difference, d, is $d = 9.4$

$a\_n = a + (n-1)d = 7 + (n-1)(9.4) + 7 + 9.4n - 9.4 = 9.4n - 2.4$

$a\_3 = 9.4(3) - 2.4 = 25.8$

In the spirit of our earlier work we look at the third term in the sequence to find a name for it because the Earth is the third planet from the sun. We find it is 25.8.

Clarifying Points

It must be clear that my communication with the Gypsy Shaman that seems real, is not real but a trick. We can assume The Gypsy Shaman has seen the movie 2001: A Space Odyssey and has designed the nature of our apparent telepathic communication according the the AE-35 antenna in that movie, the AE-35 was the communication of the ship Discover to Earth on its voyage to Jupiter. The way it works is that he we are alone together in his cave when he shows me his hose collection and has me place one in it with the others that I bought him, and at that moment he becomes open where his self is concerned, I trusting him, do the same, so he becomes me and I become him. Then he has me help him install an antenna for television reception and at that moment, our- selves again revealed to one another, I see and hear him in every thought I have, and feel even though we are separated later, he sees and hears me in every thought I have, because being him, I respond to him as he would do.

We must address why what I am doing is science. Having found nine-fifths in nature, in gold, silver, sun and moon, largest, most massive planets Jupiter and Saturn and having showed it connected profoundly to pi combined with the golden ratio and pi combined Euler's number, we have hypothesized that extraterrestrials have known this exists in nature, and interestingly in our solar system, so they would use it to send us a mes- sage. We use the relationship to find three equations that make a plane in space and with that find a normal vector to the plane, and, here is the thing, it points to Sagittarius, where the SETI Wow! signal originated from, which is the one signal so far the SETI program (Search For Extraterrestrial Intelligence) has received that has everything we would expect one to have if it was extraterrestrial. Interestingly, the signal lasts 72 sec- onds and in the movie 2001: A Space Odyssey the antenna is reported by ship comput- er HAL to fail in 72 hours. What is more 72 is a whole number multiple of the 9/5 used to make the equations that point to Sagittarius. 7.2 is four times 9/5. And, as well 72 is the amount of years it takes the Earth to precess one degree in its precession of equinoxes. As well 0.72 astronomical units is the average distance of Venus from the sun. Venus is considered a failed Earth, and the AE-35 was a failed antenna.

What I am doing is also science because we have found that even though the the weights and lengths we have assigned to our units of measurement seem to have evolved sometimes randomly, other times through a complicated history, are oddly con- nected to the occurrence of the 9/5 found in nature and the universe. We hypothesize, then, that extraterrestrials have influenced the development of our systems of units since perhaps even ancient times. We also offer the other possible explanation, that 9/5 could naturally occur in a randomly evolved system (like our systems of units) through some nature of mathematics.

Ian Beardsley
May 27, 2014

Let us look at the Calculation That Points To Sagittarius

We start with five and add 9 to each successive term:

5, 14, 23, 32,…

and take the whole number multiples of 9/5:

1.8, 3.6, 5.4, 7.2,…

We take the difference between the two sequences:

3.2, 10.4, 17.6, 24.8,…

Which is an arithmetic sequence with common difference of 7.2 meaning it is written

$7.2n - 4 = a\_n$

Start with 8 and add 5 to each additional term (we throw a twist by not starting with 5)

5/3 => 8, 13, 18, 23,…

List the numbers that are whole number multiples of 5/3:

$5/3n = 1.7$, 3.3, 5, 6.7,…

Subtract respective terms in the second sequence from those in the first:

6.3, 9.7, 13, 16.3,…

This is an arithmetic sequence with common difference 3.3. It can be written:

$(a\_n) = 3 + 3.3n$

We then sought the Yang of six-fold symmetry because it is typical to physical nature, like snowflakes. We said it was 5/3 since it represents the 120 degree measure of angles in a regular hexagon. Now we consider the alternate yang:

$360 - 60 = 300$

$300 + 360 = 660$

$660/360 = 11/6$

11/6 => 11/6, 11/3, 11/2, 22/3,... = 1.833, 3.667, 5.5, 7.333,...

11/6 => 6, 6+11 = 17, 17+11=28, 28+11=39, ... = 6, 17, 28, 39,...

Subtract the second sequence from the first:

4.167, 13.333, 22.5, 31.667,...

Now we find the common difference between terms in the latter: 9.166, 9.167, 9.167,...

$(a\_n) = a + (n-1)d = 4.167+(n-1)9.167 = 4.167 + 9.167n - 9.167 = 9.167n-5$

Try n=3: $9.167(3) - 5 = 27.501 - 5 = 22.501$ (works)

$(a\_n) = 9.167n - 5$

I have found nine-fifths occurs throughout nature in the rotation of petals around a a flower for a most popular arrangement, in the orbits of jupiter to saturn in their closest approaches to the sun, in the ratio of the molar masses of gold to silver, and in the ratio of the solar radius to the lunar orbit. I now further go on to say that this nine-fifths unifies the two most important ratios in mathematics pi and the golden ratio (phi), in that

pi + phi = 3.141 + 1.618 = 4.759

Because the numbers after the decimal in the sum (the important part) are 5 and 9 and 7, the average of 5 and nine.

I have talked about how 9/5, which I have found exists in Nature and the Universe, unifies pi and the golden ratio (phi):

(pi) + (phi) = 3.141 + 1.618 = 4.759

because the first three numbers after the decimal are 7, 5 and 9. Seven is the average of nine and five, and the second number is our 5 in nine-fifths and the third number is the 9 in nine-fifths.

It would seem 9/5 unifies euler's number, e, and pi, as well:

(pi) + (e) = 3.141 + 2.718 = 5.859

The second number after the decimal is the 5 in nine-fifths, and the third number after the decimal is the 9 in nine-fifths. The first number after the decimal is eight. This is significant because the 8 is the 8 in 1.8, which is 9/5 divided out. The one will take you all the way around a circle, what is left is 0.8.

Yin of 9/5 (five-fold symmetry)

360/5 = 72
360-72=288
288+360=648
648/360=9/5

Yang of 5/3 (six-fold symmetry)

360/6=60
360-60=300
300-60=240
240/360=2/3
2/3+1=5/3

Yang 2 of 11/6 (six-fold symmetry)

360-60=300
300+360=660
660/360=11/6

We have the Neptune Equation:

7.2x –4

We have the Uranus Equation

3.3x + 3

And now with our alternate cosmology we have The Earth Equation:

9x-5

With three equations we can write the parameterized equations in 3-dimensional space, parameterized in terms of t, for x, y, and z. We can write from that f(x,y,z) and find the gradient vector, or normal to the equation of a plane in other words, and from that a region in space.

$$x(t) = \frac{36}{5}t - 4$$

$$y(t) = \frac{33}{10}t + 3$$

$$z(t) = 9t - 5$$

$$\frac{5x+20}{36} = \frac{10y-30}{33} = \frac{z+5}{9}$$

$$\frac{5}{36}x - \frac{10}{33}y - \frac{1}{9}z + \frac{10}{11} = 0$$

$$Vf = \langle 5/36, -10/33, -1/9 \rangle$$

a=5/36   b=-10/33

$$c = \sqrt{(5/36)^2 + (10/33)^2} = \sqrt{0.0918 + 0.019} = 0.3328$$

d=-1/9

$$\tan \alpha = b/a$$

$$\alpha = -65.358°$$

$$\tan \beta = d/c$$

$$\beta = -18.46°$$

-65.358 degrees/15 degrees/hour =-4.3572 hours

24 00 00 – 4.3572 = 19.6428 hours

RA: 19h 38m 34s
Dec: -18 degrees 27 minutes 36 seconds

$$\left\langle \frac{5}{4}, -\frac{30}{11}, -1 \right\rangle \approx \left\langle 1, -3, -1 \right\rangle$$

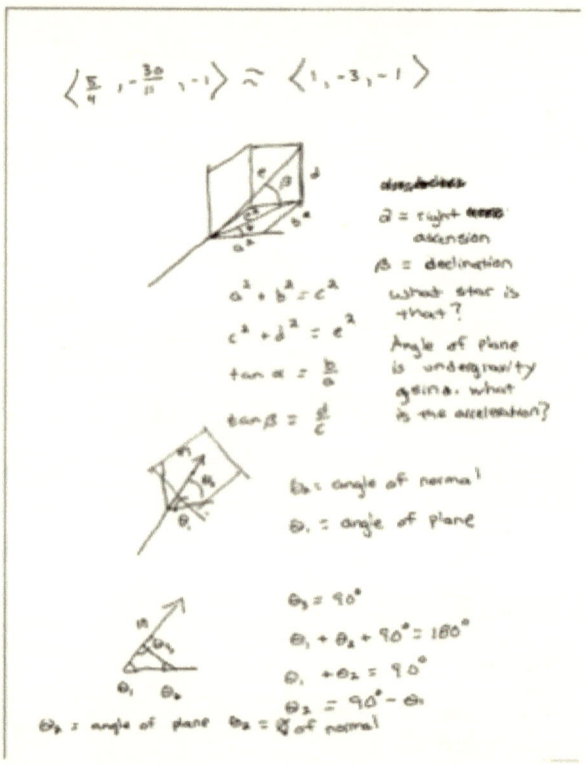

α = right ascension

β = declination

what star is that?

Angle of plane is under gravity going. what is the acceleration?

$a^2 + b^2 = c^2$

$c^2 + d^2 = e^2$

$\tan \alpha = \frac{b}{a}$

$\tan \beta = \frac{d}{c}$

θ_2 = angle of normal

θ_1 = angle of plane

$\theta_3 = 90°$

$\theta_1 + \theta_2 + 90° = 180°$

$\theta_1 + \theta_2 = 90°$

$\theta_2 = 90° - \theta_1$

θ_1 = angle of plane θ_2 = of normal

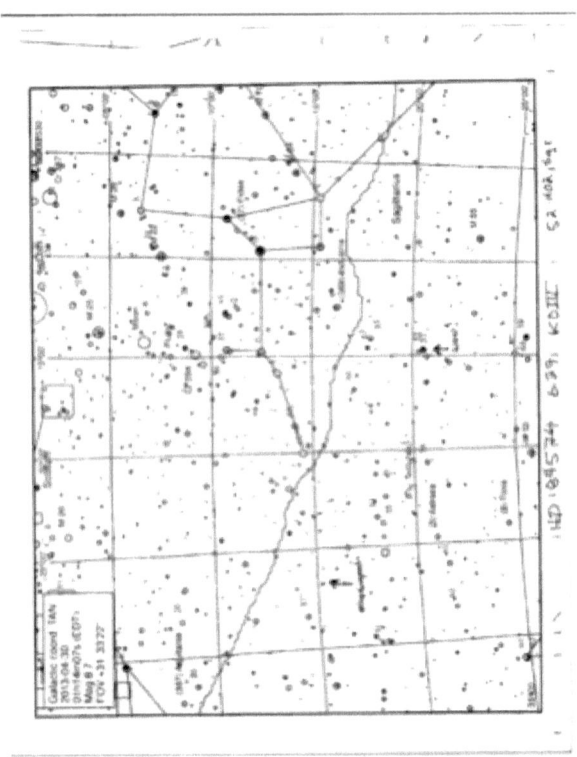

The projection by my calculation through my cosmology of yin, yang, and 15 for the origin of my message from extraterrestrials was somewhere in the easternmost part of the constellation Sagittarius. This happens to be the same place where the one possible alien signal was detected in the Search For Extraterrestrial Intelligence (SETI). It was called "The Wow Signal" because on August 15, 1977 the big ear antenna received something that seemed not like star noise, but exactly what they were looking for in an extraterrestrial signal. Its name is what it is because the astronomer on duty, Jerry R. Ehman wrote "Wow!" next to the numbers when they came in. Incredibly, it lasted the full 72 seconds that the Big Ear antenna listened for it. I say incredible because I have mentioned the importance of 72, not just in my Neptune equation – for which my location in space was derived in part – but because of its connection to the Gypsy Shaman's AE-35 antenna and its relation to 72 in the movie "2001: A Space Odyssey". The estimation of the coordinates for the origin of the Wow signal are two:

19h22m24.64s

19h25m17.01s

With declination of:

-26 Degrees 25 minutes 17.01 s

That is about 2.5 degrees from the star group Chi Sagittarri

It is very close to my calculation for an extraterrestrial civilization that I feel hid a message in our physics, which I calculate to be near HD 184835 and exactly at:

19h 38m 34s
-18 Degrees 27 minutes 36 seconds

The telescope that detected the Wow Signal was at Ohio Wesleyan University Delaware, Ohio called The Perkins Observatory.

Ian Beardsley May 6, 2013

But let us tell the story of everything that lead to this, in a short piece titled Gypsy Shamanism And the Universe.

All That Can Be Said

They originated in the Far East, and passed some through the south, and others through the north over a period of A Thousand Years through deserts. They became Artists, and camped where there was no one else, and went unseen and unknown during all that time, only to unite and settle at Land's End, in neighborhoods, barrios, or quarters of western construct, and to let out only one verse of one of their poets, who is anonymous: "If there is someone in the street, he is familiar with it. If there is someone in the street, he knows him."

Chapter 1

AE-35

I wrote a short story last night, called Gypsy Shamanism and the Universe about the AE-35 unit, which is the unit in the movie and book 2001: A Space Odyssey that HAL reports will fail and discontinue communication to Earth. I decided to read the passage dealing with the event in 2001 and HAL, the ship computer, reports it will fail in within 72 hours. Strange, because Venus is the source of 7.2 in my Neptune equation and represents failure, where Mars represents success.

Ian Beardsley
August 5, 2012

It must have been 1989 or 1990 when I took a leave of absence from The University Of Oregon, studying Spanish, Physics, and working at the state observatory in Oregon -- Pine Mountain Observatory—to pursue flamenco in Spain.

The Moors, who carved caves into the hills for residence when they were building the Alhambra Castle on the hill facing them, abandoned them before the Gypsies, or Roma, had arrived there in Granada Spain. The Gypsies were resourceful enough to stucco and tile the abandoned caves, and take them up for homes.

Living in one such cave owned by a gypsy shaman, was really not a down and out situation, as these homes had plumbing and gas cooking units that ran off bottles of propane. It was really comparable to living in a Native American adobe home in New Mexico.

Of course living in such a place came with responsibilities, and that included watering its gardens. The Shaman told me: "Water the flowers, and, when you are done, roll up the hose and put it in the cave, or it will get stolen". I had studied Castilian Spanish in college and as such a hose is "una manguera", but the Shaman called it "una goma" and goma translates as rubber. Roll up the hose and put it away when you are done with it: good advice!

So, I water the flowers, rollup the hose and put it away. The Shaman comes to the cave the next day and tells me I didn't roll up the hose and put it away, so it got stolen, and that I had to buy him a new one.

He comes by the cave a few days later, wakes me up asks me to accompany him out of The Sacromonte, to some place between there and the old Arabic city, Albaicin, to buy him a new hose.

It wasn't a far walk at all, the equivalent of a few city blocks from the caves. We get to the store, which was a counter facing the street, not one that you could enter. He says to the man behind the counter, give me 5 meters of hose. The man behind the counter pulled off five meters of hose from the spindle, and cut the hose to that length. He stated a value in pesetas, maybe 800, or so, (about eight dollars at the time) and the Shaman told me to give that amount to the man behind the counter, who was Spanish. I paid the man, and we left.

I carried the hose, and the Shaman walked along side me until we arrived at his cave where I was staying. We entered the cave stopped at the walk way between living room and kitchen, and he said: "follow me". We went through a tunnel that had about three chambers in the cave, and entered one on our right as we were heading in, and we stopped and before me was a collection of what I estimated to be fifteen rubber hoses sitting on ground. The Shaman told me to set the one I had just bought him on the floor with the others. I did, and we left the chamber, and he left the cave, and I retreated to a couch in the cave living room.

Chapter 2

Gypsies have a way of knowing things about a person, whether or not one discloses it to them in words, and The Shaman was aware that I not only worked in Astronomy, but that my work in astronomy involved knowing and doing electronics.

So, maybe a week or two after I had bought him a hose, he came to his cave where I was staying, and asked me if I would be able to install an antenna for television at an apartment where his nephew lived.

So this time I was not carrying a hose through The Sacromonte, but an antenna.

There were several of us on the patio, on a hill adjacent to the apartment of The Shaman's Nephew, installing an antenna for television reception.

Chapter 3

I am now in Southern California, at the house of my mother, it is late at night, she is a asleep, and I am about 24 years old and I decide to look out the window, east, across The Atlantic, to Spain. Immediately I see the Shaman, in his living room, where I had eaten a bowl of the Gypsy soup called Puchero, and I hear the word Antenna. I now realize when I installed the antenna, I had become one, and was receiving messages from the Shaman.

The Shaman's Children were flamenco guitarists, and I learned from them, to play the guitar. I am now playing flamenco, with instructions from the shaman to put the gypsy space program into my music. I realize I am not just any antenna, but the AE35 that malfunctioned aboard The Discovery just before it arrived at the planet Jupiter in Arthur C. Clarke's and Stanley Kubrick's "2001: A Space Odyssey". The Shaman tells me, telepathically, that this time the mission won't fail.

Chapter 4

I am watching Star Wars and see a spaceship, which is two oblong capsules flying connected in tandem. The Gypsy Shaman says to me telepathically: "Dios es una idea: son dos". I understand that to mean "God is an idea: there are two elements". So I go through life basing my life on the number two.

Chapter 5

Once one has tasted Spain, that person longs to return. I land in Madrid, Northern Spain, The Capitol. The Spaniards know my destination is Granada, Southern Spain, The Gypsy Neighborhood called The Sacromonte, the caves, and immediately recognize I am under the spell of a Gypsy Shaman, and what is more that I am The AE35 Antenna for The Gypsy Space Program. Flamenco being flamenco, the Spaniards do not undo the spell, but reprogram the instructions for me, the AE35 Antenna, so that when I arrive back in the United States, my flamenco will now state their idea of a space program. It was of course, flamenco being flamenco, an attempt to out-do the Gypsy space program.

Chapter 6

I am back in the United States and I am at the house of my mother, it is night time again, she is asleep, and I look out the window east, across the Atlantic, to Spain, and this time I do not see the living room of the gypsy shaman, but the streets of Madrid at night, and all the people, and the word Jupiter comes to mind and I am about to say of course, Jupiter, and The Spanish interrupt and say "Yes, you are right it is the largest planet in the solar system, you are right to consider it, all else will flow from it."

I know ratios, in mathematics are the most interesting subject, like pi, the ratio of the circumference of a circle to its diameter, and the golden ratio, so I consider the ratio of the orbit of Saturn (the second largest planet in the solar system) to the orbit of Jupiter at their closest approaches to The Sun, and find it is nine-fifths (nine compared to five) which divided out is one point eight (1.8).

I then proceed to the next logical step: not ratios, but proportions. A ratio is this compared to that, but a proportion is this is to that as this is to that. So the question is: Saturn is to Jupiter as what is to what? Of course the answer is as Gold is to Silver. Gold is divine; silver is next down on the list. Of course one does not compare a dozen oranges to a half dozen apples, but a dozen of one to a dozen of the other, if one wants to extract any kind of meaning. But atoms of gold and silver are not measured in dozens, but in moles. So I compared a mole of gold to a mole of silver, and I said no way, it is nine-fifths, and Saturn is indeed to Jupiter as Gold is to Silver.

I said to myself: How far does this go? The Shaman's son once told me he was in love with the moon. So I compared the radius of the sun, the distance from its center to its surface to the lunar orbital radius, the distance from the center of the earth to the center of the moon. It was Nine compared to Five again!

Chapter 7

I had found 9/5 was at the crux of the Universe, but for every yin there had to be a yang. Nine fifths was one and eight-tenths of the way around a circle. The one took you back to the beginning which left you with 8 tenths. Now go to eight tenths in the other direction, it is 72 degrees of the 360 degrees in a circle. That is the separation between petals on a five-petaled flower, a most popular arrangement. Indeed life is known to have five-fold symmetry, the physical, like snowflakes, six-fold. Do the algorithm of five-fold symmetry in reverse for six-fold symmetry, and you get the yang to the yin of nine-fifths is five-thirds.

Nine-fifths was in the elements gold to silver, Saturn to Jupiter, Sun to moon. Where was five-thirds? Salt of course. "The Salt Of The Earth" is that which is good, just read Shakespeare's "King Lear". Sodium is the metal component to table salt, Potassium is, aside from being an important fertilizer, the substitute for Sodium, as a metal component to make salt substitute. The molar mass of potassium to sodium is five to three, the yang to the yin of nine-fifths, which is gold to silver. But multiply yin with yang, that is nine-fifths with five-thirds, and you get 3, and the earth is the third planet from the sun.

I thought the crux of the universe must be the difference between nine-fifths and five-thirds. I subtracted the two and got two-fifteenths! Two compared to fifteen! I had bought the Shaman his fifteenth rubber hose, and after he made me into the AE35 Antenna one of his first transmissions to me was: "God Is An Idea: There Are Two Elements".

It is so obvious, the most abundant gas in the Earth Atmosphere is Nitrogen, chemical group 15!

The Black Night Satellite became know to the public on May 14 1954 in an article that appeared in the San Francisco Examiner. No one including NASA could explain its origins and it is thought by many to be an alien made satellite orbiting the Earth, monitoring us. It brings to mind the Monolith of 2001: A Space Odyssey. It turned out to be an alien sentinel placed on earth to monitor us and give us an evolutionary nudge when we needed it. Why did not NASA retrieve it with the space shuttle? My guess is they figured they did not know what they were dealing with. As we could learn from 2001: A Space Odyssey, the monolith should be approached with caution. The interesting thing is that A Ham Radio operator who received signals from it, decoded them and learned the satellite was 13,000 years old and came from the double star epsilon bootis. That is further interesting in that Our Cosmic Ancestors by Maurice Chatelain told of a story how there was some sort of echo in communication experiments. The echo was studied and its pattern if used on the televisions of the time came out to be a map of the constellation bootes as reported by the Scottish Astronomer Duncan Lunan, highlighting the same star that The Black Night has been reporting to as determined by the Ham Radio operator.

Back in 2005 I in my book Decoding The Universe Two I wrote something that pertains to the Black Knight Satellite. Consider that the satellite is presumed to be extraterrestrial and is sending information to the constellation Bootes. I noticed that the closest star to us, Alpha Centauri, since it is the closest and I point out now that it is composed of a star just like the Sun, that it should be connected to the Earth. Well it is: the Earth is the third planet from the Sun and Alpha Centauri is the third brightest star in the sky. I then proceed to consider the largest planet, and most massive in the Solar System and it is Jupiter. If the above pattern repeats, it should connected to the brightest star in the sky. Well it is: Jupiter is the fifth planet from the Sun and Sirius is the fifth nearest star in the sky.
Let us now consider what Carl Sagan Said: to travel in space is to sail upon the cosmic ocean. Mars is the fourth planet from the Sun and is the best candidate for a body to colonize as it is solid (not gaseous) and is not too cold, or to warm. How appropriate! The brightest star in the constellation Bootes is Arcturus and it is the fourth brightest star in the sky. Also Bootes means "The Boatman". Is the Black Knight Satellite then, since theory has it is from Bootes, guiding us towards setting sail upon the cosmic ocean to colonize Mars?

Ian Beardsley
May 2, 2014

The short and short of it is we have two sets:

(9/5, 5/3, 2)
(1, 4, 7)

They are the two vectors:

$A = \langle 9/5, 5/3, 2 \rangle$
$B = \langle 1, 4, 7 \rangle$

They lay in a plane and the normal to the plane is their cross product:

$$A \times B = \begin{array}{ccc} \bar{i} & \bar{j} & \bar{k} \\ 9/5 & 5/3 & 2 \\ 1 & 4 & 7 \end{array} = \left(\frac{5}{3}7 - 2(4)\right)\bar{i} + \left(2(1) - \frac{9}{5}7\right)\bar{j} + \left(\frac{9}{5}4 - \frac{5}{3}(1)\right)\bar{k}$$

$$= \frac{11}{3}\bar{i} - \frac{53}{5}\bar{j} + \frac{83}{15}\bar{k}$$

The magnitude of the normal vector is:

$$\left| \left(\frac{11}{3}\right)^2 + \left(\frac{53}{5}\right)^2 + \left(\frac{83}{15}\right)^2 \right| = \frac{7039}{45} = 156.4$$

$$\sqrt{156.4} = 12.5$$

$$|A \times B| = AB\sin\theta$$

$$A = \sqrt{\left(\frac{9}{5}\right)^2 + \left(\frac{5}{3}\right)^2 + 2^2} = \sqrt{3.24 + 2.78 + 4} = \sqrt{10.02} = 3.165$$

$$B = \sqrt{1^2 + 4^2 + 7^2} = \sqrt{66} = 8.124$$

$$\sin\theta = \frac{12.5}{25.71246} = 0.4861$$

$$\theta = 29°$$

We find where the normal vector to the plane that contains the vectors A and B points on the celestial sphere by converting the cross product of A and B to right ascension and declination:

$a = 11/3$

$b = -53/5$

$$c = \sqrt{\left(\frac{11}{3}\right)^2 + \left(\frac{53}{5}\right)^2} = \sqrt{13.44 + 112.36} = \sqrt{125.8} = 11.2$$

$d = \dfrac{83}{15}$

$\tan\alpha = \dfrac{b}{a} = -\dfrac{55}{158} = -0.3459$

$\alpha = -19.0805° = RA$

$\tan\beta = \dfrac{83}{15} \div 11.2 = 0.4940$

$\beta = 26.289° = dec$

$0.289(60) = 17.34$

$0.34(60) = 20.4$

$\beta = 26°17'20.4'' = dec$

$-\dfrac{19\,deg}{15\,deg/hour} = -1.267\,hours = RA$

$24\,hours - 1.267\,hours = 22.733\,hours = RA$

$(0.733)(60) = 43.98\,min$

$(0.98\,min)(60\,sec) = 58.8\,sec$

$RA = 22^h 43^m 59^s$

We now turn to our sky chart and find the brightest star closest to these coordinates is in the constellation Pegasus and is Mu Pegasus, which has the coordinates:

$22^h 50^m 42.07^s$

$24°40'32.5''$

Also, close to this star is Lambda Pegasus.

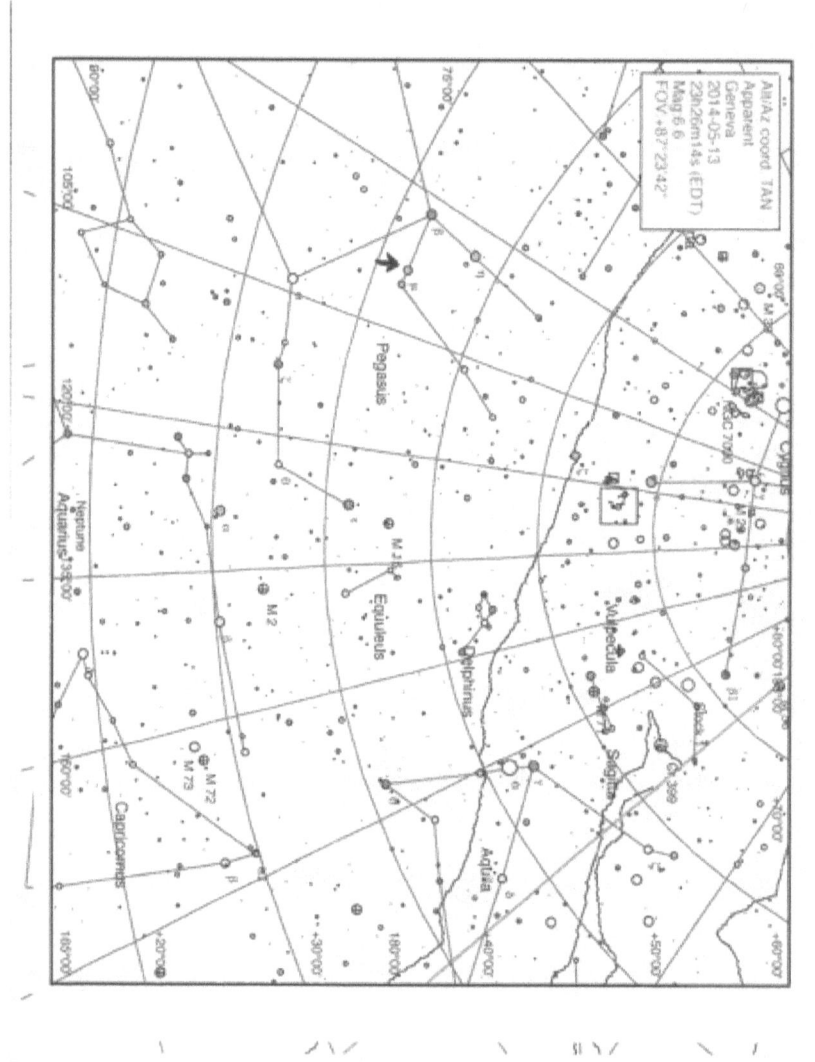

Zeta Reticuli

Bob Lazar MJ 12

No ET (out of Technology)
at area 51, they did
have the clearance. S4
was made for them.

Back engineered extraterrestrial craft.
ET Autopsy he saw the
ET looked just like in lore.
The big head, big eyes, thin
body, typical gray, could
not ascertain height from
Photo. The craft were just
like reported UFO's, disc shaped,
operate generating gravity waves.
He was told ets came from zeta
reticuli star system. Apparently some
were on the base working with them.

Lazar says
they have been with
us since
we we were
semian
and that
they have
made some
60 odd
modifications
of our genome.

when on considers
the coordinates of
Zeta Reticuli, they
find when converting
hours to degrees we
have about 45° for right ascension
and the declination is close
to 60°. These are important   60° in
angles in the special triangles   the negative direction
45 - 45 - 90, and 60-30-90.

Zeta Reticuli is a double star
system with both components
like the sun. It is only about
400 ly from us.

$\tan \frac{b}{a} = 45° = R.A$

$\tan \frac{a}{c} = 6° = $ declination

$c^2 = a^2 + b^2$

$e^2 = c^2 + d^2$

Coordinates of Zeta Reticuli

3^h 18^m Approximately

$-62°$ 18^m $45°$ and $60°$

These are the two of three
regular tessellators.

$$\tan 45° = \frac{b}{a}$$

$$\tan 60° \quad \frac{d}{c}$$

$$c^2 = a^2 + b^2$$

$$a = b$$

$$\frac{b}{a} = 1$$

$$\frac{d}{c} = 1.732 = \frac{433}{250}$$

$$d = 433 \quad c = 250$$

$$(250)^2 = 2b^2$$

$$\frac{62,500}{2} = b^2 = 31,250$$

$$b = 176.7766953 \approx 177 = a$$

The vector pointing to zeta reticuli

is $\langle a, b, d \rangle = \langle 177, 177, -433 \rangle$

Ian Beardsley
May 09, 2014

In Discover And Contact we began with Manuel, the Chovihano, or Gypsy Shaman. We began with how he made me into the AE-35 Antenna, which was the unit in the movie 2001: a Space Odyssey that was reported to fail by the ship Discovery computer HAL to astronaut Dave Bowman, within 72 hours. We noted that making me into the AE-35 antenna made me telepathic with Manuel, which was not a real telepathy, but psychological trick induced when he showed me his collection of 15 rubber hoses in his cave in Granada, Spain. 72 is important because as it would turn out the precession of the Earth Equinoxes are one degree in 72 years. The number 15 is important because the Earth rotates through 15 degrees in one hour.

But it becomes more interesting when I show that Gold to Silver in molar mass is nine to five and so is the solar radius to the lunar orbital radius, and, the sun is Gold in color, the moon silver. This results in the Neptune Equation, which has the multiplier of 7.2 from Venus Orbit in astronomical units and Venus is a failed Earth, and Hal reports the AE-35 will fail in 72 hours.

As if this was not enough, we showed that 9 to 5, which is one point eight, describes the digits after the decimals both in pi added to the golden ratio and in pi added to Euler's number and, 9/5 results in one of three equations that make a plane in space whose normal vector points to the constellation Sagittarius, where SETI, the program for the Search for Extraterrestrial Intelligence detected a signal on August 15, 1977 which had all of the characteristics of being a signal from off earth intelligence. The signal lasted exactly 72 seconds: the same 72 of the 72 hours before the AE-35 would fail aboard Discovery, the same 72 that is the 0.72 Astronomical Units of Venus, a failed Earth, from the Sun, and the same 72 years for the precession of the Earth Equinox.

7.2 occurs in my Neptune equation. Neptune is Poseidon, the Greek God of the Sea. We can say both that 2001: A Space Odyssey is a space odyssey version of the Ancient Greek work Illiad and the Odyessy by Homer, and that it is a play on the idea that when we went into space, it was not unlike the early Odysseys upon the ocean in the Homer's Illiad and the Odyssey. My Neptune equation is connected to the planet Neptune, which is named for the Greek God of the ocean: When you set sail upon the sea, you have appeal to the God of the ocean Neptune (Poseidon, Greek), and now we must appeal to Neptune the planet to travel safely through space.

Now we have delved into other experiences in the Sacromonte (Gypsy Caves), to see if we can find more. We have. We created two sets of three numbers, found their cross product and that it points to the constellation Pegasus. As it would turn out, Pegasus is the constellation where the first exoplanet was discovered to orbit a star like the Sun. Before that we looked at epsilon bootis because the constellation bootes is the boatman, and as well their seems to be some indication that there is a thirteen thousand year old satellite orbiting the earth reporting to that constellation. Finally, since government physicist, Bob Lazar, has disclosed that he was back engineering alien spacecraft and was told the craft came from the star Zeta Reticuli star system, we looked at that star. We calculated that if you convert its Right ascension to degrees, it comes out to be very close to 45 and we also note its declination is almost -60 degrees; these are important angles in special regular polygons that tesselate and and allow us to calculate trignometric functions of angles. But most interestingly, we say now, is that the constellation to which Zeta Reticuli belongs is Reticulum, which was named by an astronomer Nicolas Louis de Lacaille, to commemorate the reticle in his telescope eyepiece in the eighteenth century. He was using the eyepiece to define the constellation as a bust of Columbus. It was the Italian Columbus who was funded by Spain (Isabella) to discover a westerly route to India based on the idea that the world was round, but landed in America, discovering it.

Sometime after the Gypsy Shaman, Manuel of the Spain of Isabella, made me the AE-35 Antenna of a spaceship, not sea ship (like Columbus sailed), called Discovery, and set me on a course for contemplating space travel, I was intercepted by the Italy of Columbus, when I married a woman from Italy and went there with her several times, which set me on a course to discover that the nature of the universe and mathematical constants are connected to the central extraterrestrial activity that seems to be occurring here on Earth. Concerning my work Discover and Contact Two Cristopher Columbus died on my Birthday, May 20 and where I went in Spain was Granada, which lead to my discoveries, and the tomb of Isabella la Catolica is in Granada. She funded the voyage of Columbus.

Ian Beardsley
May 16, 2014

Contact Again

There is a way of calculating when the possible message from extraterrestrials in Sagittarius, the SETI Wow! Signal, will repeat it self. Our discovery of another message from the same place began with the Gypsy Shaman's hose collection of 15 hoses making us realize that 15 was important because the Earth rotates through 15 degrees in an hour, and 15 seconds lead to the dynamic integral, Manuel's Integral. The first message was on August 15, 1977. We noted that that 15 of August pointed to the Shaman's 15 hoses and the two sevens in 1977 add up to 14, which when added to the one in 19 is 15 as well, while the 9 is the nine of nine-fifths that we found in Nature from which we calculated a place in Sagittarius where the SETI Wow! Signal is. We decoded another message on around May 5 of 2013. The next message should then be on August 15, 2015, to line up the Shaman's fifteens. August is a good time to view the constellation Sagittarius from where I am in Southern California. Sagittarius has always been my favorite constellation, because it is in the center of the Galaxy, rich with globular, and open clusters that can be viewed with binoculars. Summer is when stargazing becomes exciting because you get both a rich sky and warm, uplifting weather. Also, August is the eighth month and our nine-fifths divided out is one point eight. The one takes you around a circle once, leaving point eight.

Dec 24, 2013

Did ETs Embed A Message In Our Physics?

A diode is made of silicon (sand) doped with phosphorous and boron, all naturally occurring elements made (forged) in interior of stars. A silicon diode needs 0.6 volts to
conduct a current. In a diode gate a high voltage is six volts and the circuit is on and a
low voltage, or ground (zero), the circuit is off. Off is coding for zero and on is coding for
1. Zeros and ones can be strung together to make numbers, and letters in the alphabet.

Manuel, the Gypsy Shaman, said: "God is an Idea, There are two elements." We can
take this to be our Yin and Yang of 9/5 and 5/3. Low voltage and the circuit is off; high
voltage and the circuit is on. These, again, are the two elements of the Shaman's yin
and yang.

Because you have to run a diode at least 0.6 volts, this is subtracted from the six volts
that turn the diode on, so really it operates on slightly less than six volts. This voltage
drop of 0.6 Volts in the circuit is what we want to look at. We will evaluate the voltage
drop over the Shaman's mysterious 15 seconds and obtain an amount of power, then
apply that power over one year – the time it takes the earth to go around the Sun. We
have already found that the Shaman's 15 not only unifies meters with the universe, but
meters with feet, amazingly in the Uranus Integral outlined both in The Exploits Of
Manuel and Further Exploits Of Manuel.

Now we do the calculation:

One electron volt is the energy of an electron falling through a potential difference of one volt, which is equal to 1.602E-19 Joules.

One Joule is 10,000,000 ergs

(0.6 Volts)(1.602E-19 Joules) = 9.612E-20 Joules

(9.612E-20 Joules)(10,000,000 ergs/Joule) = 9.612E-13 ergs

(9.612E-13ergs)/(15 seconds) = 6.408E-14 ergs/second

365.25 * 24 * 60 * 60 = 31,557,600 seconds/year

(6.408E-14 ergs/sec)(31,557,600 seconds) = 2.022211008 microergs = 2.0 microergs

It is almost exactly two microergs! (Is the two microergs prompting that we decode the message in binary?)

What Message?

My book A Thorough Dimensional Analysis is about how I provide evidence in support of the idea that extraterrestrials gave us our units of measurement, like the foot, the meter, the second, in Ancient times despite the history that tells us the metric system is modern, and the foot-pound system, perhaps a little ambiguous in its history. I think along the way in doing this I may have found an encrypted message from extraterrestrials in the number:

2.022211008

Insofar as it came to us from natural constants like the charge of an electron, the orbital period of the Earth, and numbers we invented like the measure of electromotive potential, the volt. The idea is that extraterrestrials somehow told us to measure electromotive potential with a unit of volts because it would produce the above number. I am interested in it because of the repetition of ones and twos, and the separation of numbers by zeros.

Let us look at the part after the decimal. It is:

022211008

It is the second number three times, the first number two times and the last number is the number of digits that precede it.

The three twos and the two ones add up to the last digit, eight.

The two ones add up to each of the three twos.

It counts from right to left 0,1,2,... .

Decoding The Message

The message is:

022211008

The 222 adds up to six, which in binary is 110. The next segment is 1100, which is a nibble, and is the number 12. This segment is in binary to begin with. The last digit is an eight, which is 1000 in binary. Thus we have, including the first digit, a zero, which is a bit:

(0) (110) (1100) (1000)

We write this as three nibbles:

(0110) (1100) (1000)

Thus 022211008 translates as 0110 1100 1000

Which are the three numbers:

6, 12, 8

These three numbers divide nicely into the 360 degrees of a circle:

360/12 = 30   360/6 = 60   360/8 = 45

These are the angles in the three special triangles with which we can write out the trigonometric functions of important angles in closed form:

30-60-90   60-60-60   45-45-90

6, 12, 8 have the common denominator of two. This gives us the three numbers 3, 6, 4:

364

Which is very close to the number of Earth days in a year (365). Are extraterrestrials pointing out that the number of Earth days in a year are closely connected to the three special triangles? The number in binary is interesting, because the ones scroll off from right to left nibble to nibble, one digit at a time. The message is also that the number of days in a year are connected to the three regular tessellators. They have sides 3, 6, and 4.

Did ETs Embed A Message In Our Computer Science?

I have said, since my theory suggest extraterrestrials gave us our units of measurement, that extraterrestrials might have given us our variables used in physics and math, like the unit vectors (i, j, k). I have already found a pattern and posted it. However, I was doing my CS50x computer science homework and trying to write a program for Caesar's Cipher. I wrote a small program and decided to test it. If you write a program and test it, standard input is "hello". I put in hello and to test, ran the program for rotating characters by 1, and 2, and 3, as they are the first integers and the easiest with which to test your program. The result was the "h" on "hello", came out to be (i, j, k). In other words you get that (i, j, k) is a hello from aliens in accordance with my earlier theories. If this is not real contact with extraterrestrials, it is great content for a Sci-Fi movie about contact with extraterrestrials. Here is the program I wrote, and the result of running it:

As you can see I am making some kind of a cipher, but not Caesar's Cipher

```
#include <stdio.h>
#include <cs50.h>
#include <string.h>
int main(int argc, string argv[1])
{
int i=0;
int k = atoi(argv[1]);
if (argc>2 || argc<2)
printf ("Give me a single string: ");
else
printf("Give me a phrase: ");
string s = GetString();
for (int i =0, n=strlen(s); i<n; i++);
printf("%c", s[i]+k);
printf("\n");
}
```

Running Julius 01

```
jharvard@appliance (~): cd Dropbox/pset2
jharvard@appliance (~/Dropbox/pset2): make julius
clang -ggdb3 -O0 -std=c99 -Wall -Werror   julius.c -lcs50 -lm -o julius
jharvard@appliance (~/Dropbox/pset2): ./julius 3
Give me a phrase: hello
k
jharvard@appliance (~/Dropbox/pset2): ./julius 4
Give me a phrase: hello
l
jharvard@appliance (~/Dropbox/pset2): ./julius 2
```

```
Give me a phrase: hello
j
jharvard@appliance (~/Dropbox/pset2): ./julius 1
Give me a phrase: hello
i
jharvard@appliance (~/Dropbox/pset2):
```

I posted to my blog http://cosasbiendichas.blogspot.com/

Sunday, January 26, 2014
A Pattern Emerges

(a, b, c) in ASCII computer code is (97, 98, 99) the first three numbers before a hundred and 100 is totality (100%).

(i, j, k) in numeric are is (9, 10, 11) the first three numbers before twelve and 12 is totality in the sense that 12 is the most abundant number for its size
(divisible by 1,2, 3, 4, 6 = 16) is larger than 12).

(x, y, z) in ASCII computer code is (120, 121, 122) the first three numbers before 123 and 123 is the number with the digits 1, 2, 3 which are the numeric numbers for the (a, b, c) that we started with.

Thursday, January 23, 2014
We Look Further Into Human Definitions That Seem Arbitrary

Just as we found our units of measurement, what they evolved into being and how we defined them, are centered around the triad of 9/5, 5/3, and 15, we might ask are our common usage of variables connected to Nature and the Universe as well. In pursuing such a question we look at:

(x, y, z,) as they represent the three axis is rectangular coordinates. We look at (i, j, k) as as they are the representations for the unit vectors, and they correspond respectively to
(x, y, z). We also look at (n) as it often means "number" and we look at (p and q) as they range from 0 to 1, in probability problems. We might first look at their binary and hexadecimal equivalents to get a start, if not their decimal equivalents. (i) is also often "integer" and (a, b, c) are the coefficients of a quadratic and are the corners of a triangle. We might add that (s) is length, as in physics dW=F ds. (a, b, c) have the same kind of correspondence with (x, y, z) as (i, j, k). All three sets, then, line up with one another and are at the basis of math and physics.

Ian Beardsley studied physics at The University of Oregon and worked in astronomy at Pine Mountain Observatory in the high desert east of Bend, Oregon for four years. His name appears on several papers in The Astrophysical Journal.

www.ingramcontent.com/pod-product-compliance
Lightning Source LLC
Chambersburg PA
CBHW021854170526
45157CB00006B/2444